Getting Started with .NET Gadgeteer

Simon Monk

O'REILLY®

Beijing · Cambridge · Farnham · Köln · Sebastopol · Tokyo

Getting Started with .NET Gadgeteer
by Simon Monk

Editor: Rachel Roumeliotis
Production Editor: Melanie Yarbrough
Cover Designer: Karen Montgomery
Interior Designer: David Futato
Illustrator: Rebecca Demarest

May 2012:   First Edition.

Revision History for the First Edition:
    May 04, 2012   First release
See *http://oreilly.com/catalog/errata.csp?isbn=9781449328238* for release details.

ISBN: 978-1-449-32823-8
[LSI]
1336143521

Contents

Preface v

1/Getting Started with .NET Gadgeteer 1
  What Is .NET Gadgeteer? 1
  The Fez Starter Kit 2
  Powering Up 4
  Installation 5
    Mac and Linux Users 5
    Install Visual Studio Express 7
    Install .NET Micro Framework 8
    Install the .NET Gadgeteer Package 10
    Upgrading the Firmware 11
  Hello World 13
    Creating a New Project 14
    Adding Components 15
    Trying It for Real 19
  Conclusion 20

2/Spy Camera 21
  The Design 22
  The Camera Module 22
    Trying It Out 27
  WPFWindow 27
    Creating a UI Structure 27
    Showing Messages 29
  The Final Program 30
  Touch Screen 30
  Writing to SD cards 32
  Testing 33
  Conclusion 33

3/Snowflakes Game 35
  The Design 35
  Screen Layout 35
  Using the Joystick module 40
  Conclusion 43

4/Web Messenger **45**

Web Servers **46**

A Gadgeteer Web Server **47**

Testing **52**

A Gadgeteer Sketch Pad **53**

Putting It All Together **57**

Testing **58**

Sharing with the World **59**

Conclusion **60**

5/Camera Backup Gadget **61**

The Design **62**

The USB Host Module **63**

The Program **64**

Copying Files **65**

Testing **68**

Breakpoints **68**

Conclusion **70**

6/What Next? **71**

Other Modules **71**

Environmental Sensors **71**

Pulse Oximeter **72**

Cellular Radio **73**

Relay Module **73**

Music **73**

WiFi **74**

Physical Design Files **74**

Documentation **75**

Blogs and Web Resources **76**

Conclusion **76**

Preface

Introduction

.NET Gadgeteer is a wonderful system for constructing prototypes of electronic gadgets. It is supported by a range of modules that are connected to a Mainboard with plug-in cables.

This book will lead you through the process of installing the necessary software onto your computer and then take you through a series of exciting projects that will give you the confidence and skill to start designing your own gadgets.

What You Will Need

The book restricts itself to using components from the Fez Starter Kit developed by GHI Electronics.

This particular kit is probably the most popular starter kit and is available from a range of vendors on the Internet, including GHI themselves (*http://www.ghielectronics.com*).

All the programs for the projects can be downloaded from the book's website at *http://www.gadgeteerbook.com*.

How to Use this Book

You need to read Chapter 1 to get you started, but then you can pick and choose from the remaining project chapters.

Conventions Used in This Book

The following typographical conventions are used in this book:

Italic
Indicates new terms, URLs, email addresses, filenames, and file extensions.

Constant width

Used for program listings, as well as within paragraphs to refer to program elements such as variable or function names, databases, data types, environment variables, statements, and keywords.

Constant width bold

Shows commands or other text that should be typed literally by the user.

Constant width italic

Shows text that should be replaced with user-supplied values or by values determined by context.

 TIP: This icon signifies a tip, suggestion, or general note.

 CAUTION: This icon indicates a warning or caution.

Using Code Examples

This book is here to help you get your job done. In general, you may use the code in this book in your programs and documentation. You do not need to contact us for permission unless you're reproducing a significant portion of the code. For example, writing a program that uses several chunks of code from this book does not require permission. Selling or distributing a CD-ROM of examples from O'Reilly books does require permission. Answering a question by citing this book and quoting example code does not require permission. Incorporating a significant amount of example code from this book into your product's documentation does require permission.

We appreciate, but do not require, attribution. An attribution usually includes the title, author, publisher, and ISBN. For example: "*Getting Started with .NET Gadgeteer* by Simon Monk (O'Reilly). Copyright 2012 O'Reilly Media, Inc., 978-1-449-32823-8."

If you feel your use of code examples falls outside fair use or the permission given above, feel free to contact us at *permissions@oreilly.com*.

Safari® Books Online

Safari Safari Books Online (*www.safaribooksonline.com*) is an on-demand digital library that delivers expert content in both book and video form from the world's leading authors in technology and business.

Technology professionals, software developers, web designers, and business and creative professionals use Safari Books Online as their primary resource for research, problem solving, learning, and certification training.

Safari Books Online offers a range of product mixes and pricing programs for organizations, government agencies, and individuals. Subscribers have access to thousands of books, training videos, and prepublication manuscripts in one fully searchable database from publishers like O'Reilly Media, Prentice Hall Professional, Addison-Wesley Professional, Microsoft Press, Sams, Que, Peachpit Press, Focal Press, Cisco Press, John Wiley & Sons, Syngress, Morgan Kaufmann, IBM Redbooks, Packt, Adobe Press, FT Press, Apress, Manning, New Riders, McGraw-Hill, Jones & Bartlett, Course Technology, and dozens more. For more information about Safari Books Online, please visit us online.

How to Contact Us

You can write to us at:

Maker Media
1005 Gravenstein Highway North
Sebastopol, CA 95472
800-998-9938 (in the United States or Canada)
707-829-0515 (international or local)
707-829-0104 (fax)

Maker Media is a division of O'Reilly Media devoted entirely to the growing community of resourceful people who believe that if you can imagine it, you can make it. Consisting of Make magazine, Craft magazine, Maker Faire, as well as the Hacks, Make:Projects, and DIY Science book series, Maker Media encourages the Do-It-Yourself mentality by providing creative inspiration and instruction.

For more information about Maker Media, visit us online:

MAKE: *www.makezine.com*
CRAFT: *www.craftzine.com*
Maker Faire: *www.makerfaire.com*
Hacks: *www.hackszine.com*

We have a web page for this book, where we list examples, errata, and plans for future editions. You can find this page at *http://oreil.ly/getstarted_dot netgadgeteer*.

To comment on the book, send email to *bookquestions@oreilly.com*.

For more information about our books, courses, conferences, and news, see our website at *http://www.oreilly.com*.

Find us on Facebook: *http://facebook.com/oreilly*

Follow us on Twitter: *http://twitter.com/oreillymedia*

Watch us on YouTube: *http://www.youtube.com/oreillymedia*

Acknowledgements

I thank Linda for giving me the time, space, and support to write this book and for putting up with the various messes my projects create around the house.

Thanks to Patrick Olivier, Tom Bartindale, Thomas Smith and Jonathan Hook from Culture Lab, for their excellent feedback and help. Many thanks to the technical reviewers, Scott Gowell and William Wallace.

Finally, I would like to thank Rachel Roumeliotis, Brian Jepson and everyone an O'Reilly who has had a hand in producing this book.

1/Getting Started with .NET Gadgeteer

If you like making things, then you will love .NET Gadgeteer. You can create many interesting projects by just plugging together components and writing a few lines of programming code.

What Is .NET Gadgeteer?

The .NET Gadgeteer system is developed by Microsoft to allow people to build small electronic projects. The modules that are available all conform to the .NET Gadgeteer standard, but are created by various different manufacturers. The most important of these components is the Mainboard. This is the board that contains the processor chip that will drive all the other components that we attach to it.

One of the most popular Gadgeteer systems is the FEZ Spider (shown in Figure 1-1), which is available as part of a kit of components with everything you need to get started. The kit includes a mainboard, a camera, multicolor LEDs, a color touch screen, a joystick, and various means of connecting to other computers and the Internet.

To build the projects in this book, you will need to buy one of these kits, which are available from a number of different sources. Just type "Fez Spider Starter Kit" into your favorite search engine.

All the components of the Starter Kit are plugged together using short leads with connectors on each end. These are supplied as part of the kit. This means that the projects can be constructed without the need for any soldering.

We can plug together hardware as much as we like, but it will not do anything until we have programmed the Mainboard. The Mainboard can be thought of as the brain of the system and all the other components are attached to the mass of connectors on the top of the board. And, with everything plugged in, it does resemble a spider of the kind of dimensions normally found only in South American jungles.

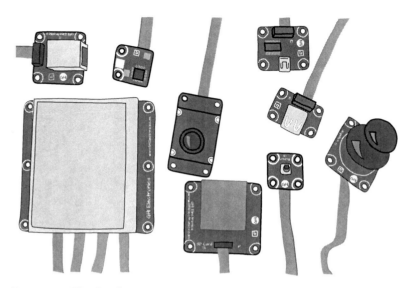

Figure 1-1. *The Fez Spider Starter Kit*

You program a Gadgeteer Mainboard by attaching it to your computer using a USB lead and running Microsoft's Integrated Development Environment called Visual Studio C# Express. One of the really nice features of Gadgeteer is that Visual Studio will let you do a lot of the programming just by drawing diagrams (see Figure 1-2).

The Fez Starter Kit

When you get your Fez Starter kit, you will find lots of little bags containing all the various components supplied with the kit. Spend some time taking each one out and figuring out what it is.

The contents of the kit may change as new versions are produced, you will need the parts outlined in Table 1-1 to build the projects in this book.

Table 1-1. *Fez Starter Kit*

Part	Description
FEZ Spider Mainboard	
Display T35	3.5" with touchscreen
USB Client DP Module	Supplies power and USB for programming
Camera Module	Low resolution webcam style camera

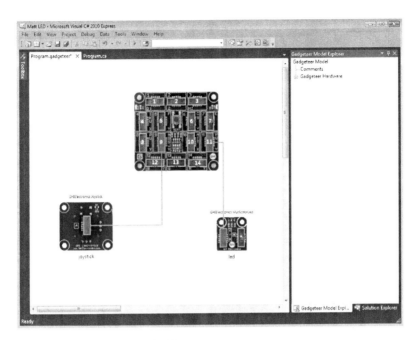

Figure 1-2. *Visual Studio C# Express*

Part	Description
2x Multicolor LED	RGB (Red Green Blue) LED can be set to any color
2x Button Module	Simple push button and independent indicator LED
Ethernet J11D Module	Network Interface
SD Card Module	For reading and writing to SD cards
USB Host Module	Attach keyboard, mouse, or USB storage device
Extender Module	Allows external components to be attached, such as temperature sensors
Joystick Module	2 axis joystick with press button

Some parts, like the LCD Display, are easy to identify while some, like "the extender," are less obvious. If you look closely, they all have a label printed on them. Most of the modules have a single connector on them that will be attached to a lead that then plugs into one of the connectors on the

Mainboard. Some of the components, like the display, have more than one connector.

We will use all of these components in the projects contained in this book.

Powering Up

Before we get started, we should check that our board isn't dead on arrival.

Identify the Mainboard, the "USB Client DP" module, and the USB lead (shown in Figure 1-3). Attach one end of a lead to the connector on the USB Client DP module and the other into the connector labelled "1" on the Mainboard. Note that the connectors have a little notch in one side so that they will connect only one way around.

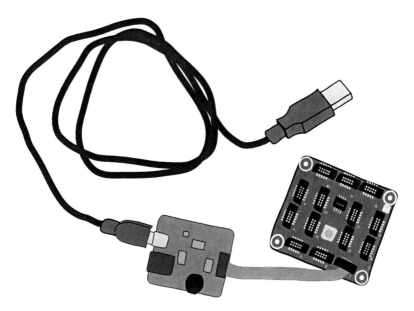

Figure 1-3. *Mainboard and USB Client DP Module*

The USB Client DP module serves to supply power to the Mainboard and hence any other components attached to it. It also attaches the Mainboard to your computer via a USB lead so that programs can be loaded onto it.

If you plug the USB lead into your computer, you should see that LEDs light on both the Power and Mainboards (see Figure 1-4).

For the time being, ignore any messages about drivers coming from your computer; we will come to that in the next section.

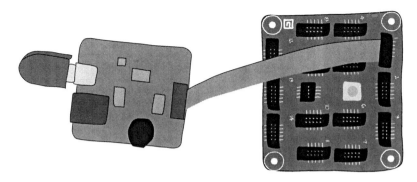

Figure 1-4. *Mainboard Powered Up*

Installation

To use Gadgeteer, we need a computer; it must be a computer that can run Windows XP, Vista, or 7. If you are a Mac or Linux user, then all is not lost if you have virtualization software like VirtualBox, which is free, you will be able to use Visual Studio Express on a Windows virtual machine.

You will, however, need a license for any version of Windows that you are running on your virtual machine.

There are four steps involved in getting your computer set up. Each is described in the following sections. The basic steps are:

- Install Visual Studio C# Express
- Install .NET Micro Framework 4.1 SDK
- Install GHI NETMF v4.1 and .NET Gadgeteer Package
- Upgrade the Mainboard's firmware

Shortcuts to all these links can be found at the Tinyclr website (*http://www .tinyclr.com/support*).

Mac and Linux Users

If you are using a Windows computer, skip this section.

If you are a Mac or Linux user, create a Windows virtual machine, being generous in memory allocation as Visual Studio Express is quite demanding of resources.

These instructions are for VirtualBox, but they will be similar for other virtualization software.

You will also need to allow the USB connection to the Gadgeteer to pass through to the virtual machine without OS X or Linux interfering with it. To do this, select your virtual machine in the VirtualBox control window and click Settings. Select Ports and then USB, as shown in Figure 1-5. Create an open USB filter that automatically forwards all USB traffic to the virtual machine. You can do this by clicking on the orange dot in the filter list.

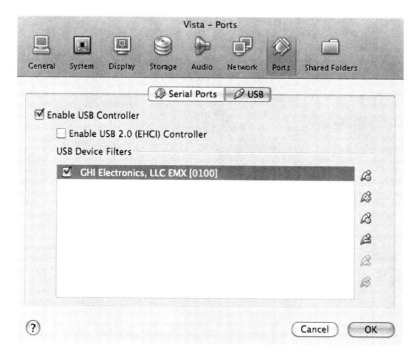

Figure 1-5. *Port Forwarding on Virtual Box*

Once you have updated the firmware on your Mainboard, you can return here and add the specific device filter by clicking the green "Add connection" icon on the right and selecting "GHI Electronics, LLC EMX." You can then remove the open filter.

Unplug your Gadgeteer from the computer until we are ready to install the USB drivers.

The rest of the steps in the installation are just carried out from your Windows virtual machine.

Install Visual Studio Express

Visual Studio Express is not a small piece of software; the installer warns you that you will need 2.4GB of disk space. You will also need a little patience.

Open your browser on *www.tinyclr.com/support*. TinyCLR is managed by GHI Electronics, which makes the Fez Starter board.

Click the "Microsoft Visual C# Express 2010" link. Then click the "Install Now" button. The basic version of this software is free (with some license restrictions). You do not need to buy the professional version. You may find that the free express version is not made quite as prominent as the professional version on Microsoft's website. In time, you may find that you want to upgrade to a paid version to take advantage of its added features, but this is not necessary for the projects in this book.

When the installer runs, you will be prompted to either Run or Save. Select Run.

 NOTE: If you are installing on Windows XP, then you must have at least Service Pack 3 installed.

You may also receive the prompt: "Program Needs your Permission to Continue." Click Continue.

The Welcome to Setup screen will appear (see Figure 1-6). Click "Next" and then accept the license agreement.

If the installer asks you if you want to install Silverlight or SQLServer, you can do so if you want, but neither are necessary for Gadgeteer use.

You can accept the default installation directory, which is *C:\Program Files \Microsoft Visual Studio 10.0*. You may have a newer version number on the end than this.

There will then be a delay while all the necessary files are downloaded and installed (see Figure 1-7). This will probably take at least 10 minutes and may be a lot longer depending on the speed of your network and computer.

During the installation, the installer may restart your computer and then continue.

Figure 1-6. *Visual Studio C# Express Installer Setup Screen*

Install .NET Micro Framework

Now we can turn our attention to installing the .NET Micro Framework, which is another Microsoft product. This is the second link on the TinyClr download page labelled "Microsoft .NET Micro Framework 4.1 SDK."

This will prompt you to Open or Save. Select Open, which will download an executable zip file.

When the zip file opens, extract all the files to the desktop, by clicking Extract All Files (see Figure 1-8).

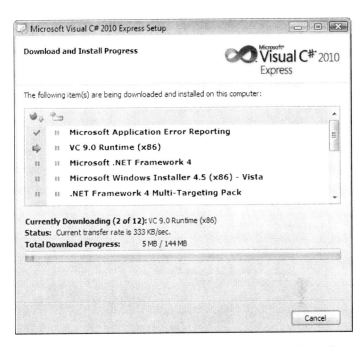

Figure 1-7. *Visual Studio C# Express Installer Download and Install*

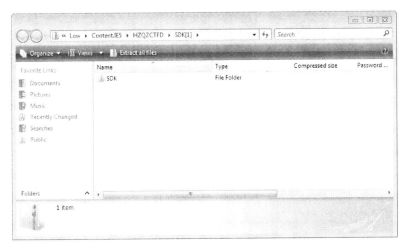

Figure 1-8. *NET Micro Framework Installation*

Inside the extracted folder, open the file *MicroFrameworkSDK*, which will run the installer (see Figure 1-9).

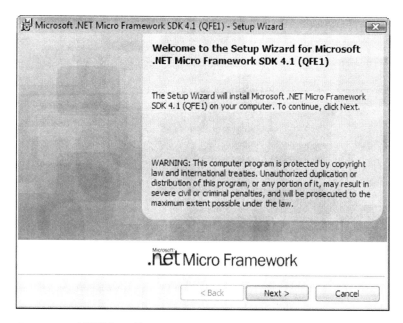

Figure 1-9. *NET Micro Framework Installer*

Accept the license agreement and select the typical install option, which should be the default.

Once the installation is complete, you can delete the zip file and installer from your desktop if you wish.

At the end of the installation, a text file showing the release notes will be displayed. You can just close this.

Install the .NET Gadgeteer Package

Open your browser on the TinyClr downloads page again (*http://www.ti nyclr.com/support/*) and look for the link to "Install the .NET Gadgeteer Package." Click the link to start the download and then select Open as the file download starts.

This will again open a zip file that you should extract onto the desktop.

Open the extracted folder and run the "Setup" application (Figure 1-10). This all-in-one installer will not only install the files you need for the Fez Spider from GHI but also the software needed for the .NET Gadgeteer modules provided by the company SeeedStudio. It will also install the USB drivers required by the hardware.

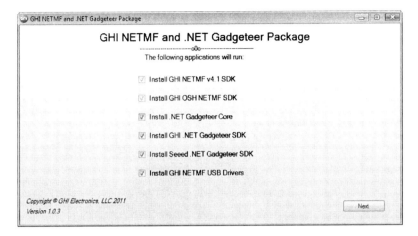

The following applications will run:

☐ Install GHI NETMF v4.1 SDK

☐ Install GHI OSH NETMF SDK

☑ Install .NET Gadgeteer Core

☑ Install GHI .NET Gadgeteer SDK

☑ Install Seeed .NET Gadgeteer SDK

☑ Install GHI NETMF USB Drivers

Copyright © GHI Electronics, LLC 2011
Version 1.0.3

Figure 1-10. *NET Gadgeteer Installer*

Accept all the defaults of the installers that will each be run automatically in turn. There are six installers in total.

Upgrading the Firmware

Once the drivers have been installed, you will be prompted to upgrade the firmware of your Mainboard. In its flash memory, the Mainboard has some of the code that is used in a .NET Gadgeteer project. By doing a firmware upgrade, we will ensure that however long the Mainboard has been sitting around before you bought it, you will still have the latest version of the firmware installed. Your board will work without making a firmware upgrade, but you may get unexpected problems as any bug fixes or improvements that have been made since the board was manufactured will not be on the board. So it is worth upgrading the firmware at least for this first installation.

WARNING: Take great care upgrading the Firmware and DO NOT unplug the board while an upgrade is in progress.

Figure 1-11 shows the first of the screens for updating the firmware. Select the Mainboard that you have from the list and click Update.

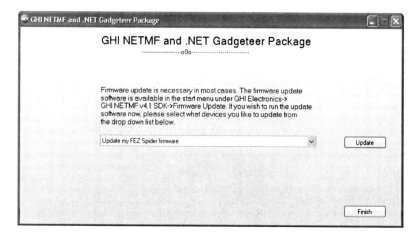

Figure 1-11. *Upgrading the Firmware—Selecting the Mainboard*

Follow the instructions in Figure 1-12. Unless you changed the installation location when you were running the installer, you can accept the default locations for the files. Click Step 2.

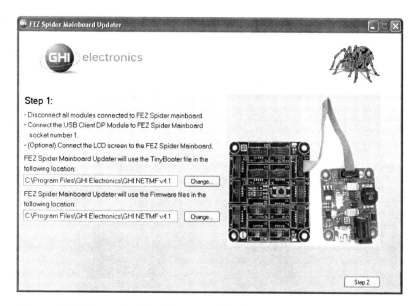

Figure 1-12. *Upgrading the Firmware—File Locations*

The final step involves moving the tiny little switches on your Mainboard as shown in Figure 1-13. This is best done with a small screwdriver. The left-most switch can stay in the up position.

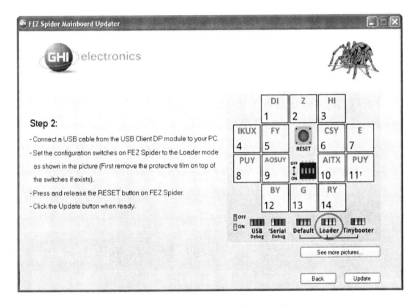

Figure 1-13. *Upgrading the Firmware—Setting the Switches*

Once the firmware has been updated, remember to unplug the board and move all the switches back to their original positions.

We are now ready to make sure everything is installed correctly by running the traditional "Hello World" project.

Hello World

In the world of programming, "Hello World" is the traditional first program that anyone writes when learning a new programming language. In the case of Gadgeteer, where we are concerned with hardware, we will make our first program turn an LED on and off when we press a button.

Start Visual Studio by launching the "Microsoft Visual C# 2010 Express" shortcut in your Start menu, which will be found inside All Programs→ Microsoft Visual C# 2010 Express.

Creating a New Project

When Visual Studio has started up, you will see a welcome screen with the options "New Project..." and "Open Project...". Select New Project; the new project screen will appear as shown in Figure 1-14.

Figure 1-14. *Visual Studio New Project Screen*

On the left-hand side, you will see a folding tree view where you can select the type of project that you want to create. Visual Studio is a general purpose integrated development platform, so we have to specify that we want a Gadgeteer project by selecting the Gadgeteer option within Visual C#.

We also need to provide the project with a name, which is done at the bottom of the window. We can just call the project "Hello". When you have done this, click OK, and Visual Studio will construct the basic framework of a new project for us. When it has finished, Visual Studio will look something like Figure 1-15.

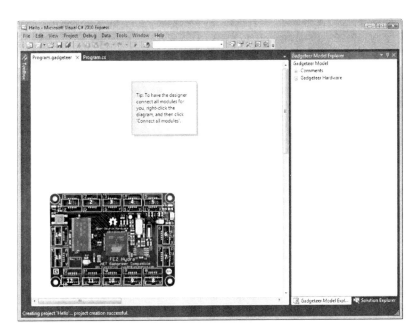

Figure 1-15. *A New Project*

Adding Components

This view of Visual Studio will allow us to add to the canvas the components that we want to use in our project and join them up to the Mainboard. In this case, we are going to use a button and a multicolor LED, as well as our Mainboard and USB Client DP module. We are using the USB Client DP module to program the Mainboard and provide power to it, but we do not need to add this to the canvas.

First, let's delete the yellow note, as we do not need this. Right-click on it and select Delete. Our setup lets us use a number of different Mainboards. If the Mainboard on the screen is a Fez Spider and that is what we are using, then great, but we may find that the new project is given a different board, such as the Fez Hydra, which has more connections.

If you do not have a Fez Spider on your canvas, but rather have some other Mainboard, then delete it in the same way you did the note. You will then need to add a Fez Spider Mainboard by clicking the Toolbox on the left. A list of components will appear. Scroll down to the "GHI Electronics" section and then drag a Fez Spider onto the canvas (see Figure 1-16).

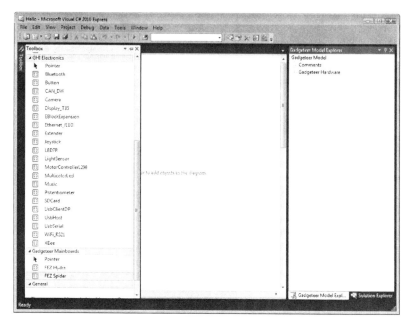

Figure 1-16. *The Toolbox*

Now drag both a Button and a Multicolor LED onto the canvas.

There are various types of connectors on the board, and not all are suitable for connecting all components to, so if you click the connector of the button that you have just dragged onto the canvas, you will see all of the compatible connectors on the Mainboard highlighted. Click again on connector 4 on the Mainboard.

Connect the left-hand connector on the Multicolor LED to connector 5 of the Mainboard. Your canvas should now look like Figure 1-17. The LED module has two connectors so that you can daisy-chain a number of LED modules to each other, without using another Mainboard connector.

We should now save what we have done by clicking the File menu and selecting Save All. This will open the Save Project dialog box. If you like, you can group a number of projects into a "Solution," however, for a Gadgeteer project it usually makes sense to just think in terms of individual projects, so we can just accept the defaults (shown in Figure 1-18).

Now we need to do a little bit of conventional programming so that when our button is pressed, it can toggle the LED on and off. Switch to the tab labeled "Program.cs" (see Figure 1-19).

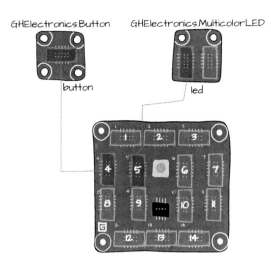

GHElectronics.Button   GHElectronics.MulticolorLED

button   led

Figure 1-17. *Adding a button and LED*

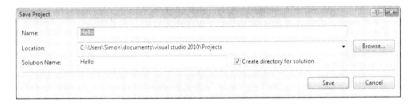

Figure 1-18. *Saving the Project*

This code has been generated for us by Visual Studio. It mostly consists of a helpful comment telling us what we need to do to have our project do something useful.

This comment is particularly useful:

```
Many modules generate useful events. Type +=<tab><tab> to add a handler to
an event, e.g.:
button.ButtonPressed +=<tab><tab>
```

It tells us how to run some code when a button is pressed. This is exactly what we want.

Visual Studio has a feature called auto-completion. This means that as we start to type some code, it will make a good guess at what we are trying to do and type a lot of the code for us.

So, just after the line:

```
Debug.Print("Program Started");
```

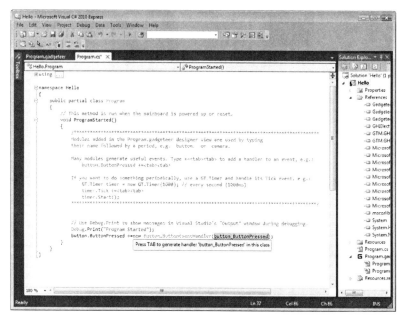

Figure 1-19. *The Code*

type:

```
button.ButtonPressed +=
```

Then press the Tab key twice. The first time you press the Tab key, it will add the text:

```
new Button.ButtonEventHandler(button_ButtonPressed);
```

The second time, it will add a whole new chunk of code:

```
void button_ButtonPressed(Button sender, Button.ButtonState state)
{
    throw new NotImplementedException();
}
```

If you are new to programming, this is called a method. It groups together a sequence of instructions to be performed. This method will be invoked every time we press the button, so it is here that we need to change the state of the LED.

Modify the **button_ButtonPressed** method so that it looks like this:

```
void button_ButtonPressed(Button sender, Button.ButtonState state)
{
    if (led.GetCurrentColor() == GT.Color.Black)
```

```
    {
        led.TurnRed();
    }
    else
    {
        led.TurnOff();
    }
}
```

This code checks to see if the current color of the LED is black (off) and if it is, it sets the color to red, otherwise it turns the LED off.

Trying It for Real

Finally, we are ready to try deploying the project to the real components.

Using the canvas shown in Figure 1-17 as a guide, connect the real components to the Mainboard as well as the USB Client DP module. When they are all connected together (see Figure 1-20), plug the USB cable into your computer.

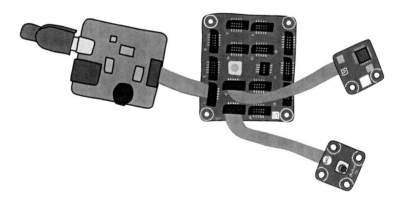

Figure 1-20. *The Hello World Project*

On the top row of your Visual Studio toolbar, you will notice a little blue triangle Play button. This will deploy our program to the Fez Spider Mainboard.

When you press the button, the status area at the bottom of the window will tell you what is going on. This should finish with a "Ready" status.

Try pressing the button. If all is well, this will turn the LED red. Pressing it again will turn the LED off.

If you modify your code now, you will find that it can't be modified. This is because you are in something called the Debugger. The Debugger is a useful tool for sorting out programs that are not behaving as expected. It allows you to set breakpoints to interrupt the code and see what is happening.

We will return to the debugger in a later chapter, but for now, if we want to exit the debugger and get back to the main code editor, then we can click on the small blue stop button on the second row of the toolbar.

Conclusion

Congratulations, you have just written and deployed your first .NET Gadgeteer program.

In Chapter 2, we will start on a simple project involving the camera and display.

2/Spy Camera

This project (shown in Figure 2-1) uses the camera module to capture images periodically and save them to the SD card.

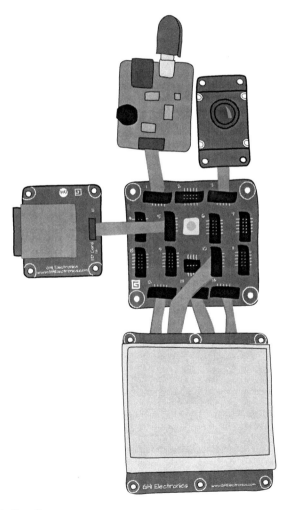

Figure 2-1. *Spy Camera*

The LCD touchscreen is used to set the recording interval of between 5 and 60 seconds.

The Design

Table 2-1. *What you will need*

Item	Source
Test Project for Camera	*http://www.gadgeteerbook.com/ downloads/2.1.SpyCameraTest.zip*
Final Full Project	*http://www.gadgeteerbook.com/ downloads/2.2.SpyCamera.zip*
Fez Spider Mainboard	*http://www.ghielectronics.com/cata log/product/269*
USB Client DP module	*http://www.ghielectronics.com/cata log/product/280*
Camera Module	*http://www.ghielectronics.com/cata log/product/283*
Display T35 Module	*http://www.ghielectronics.com/cata log/product/276*
SD Card Module	*http://www.ghielectronics.com/cata log/product/271*
SD Card	

The components are all included in the Fez Starter Kit.

Once you have downloaded the zip files for the projects from the book's website (*http://www.gadgeteerbook.com/downloads*) you will need to un-zip them into your Gadgeteer's projects area, which is usually *My Documents \Visual Studio 2010\Projects*. You can then open the project from the File menu by selecting "Open Project.." from Visual Studio's menu.

The Camera Module

Rather than jump straight to the full program for this project, we are going to start with a very simple example then build it up. This will become the "Spy Camera Test" project. You can download this later, but for now, it will be a good exercise to build it from scratch as described below.

Create a new project and give it the name "Spy Camera Test." We have added "Test" to its name so as not to get it mixed up with the final Spy Camera

Project. If you are not sure how to make a new project, refer to "Creating a New Project" on page 14.

As with our "Hello World" project in Chapter 1, the first step in this project is to add the necessary components to the design in Visual Studio's designer. Make sure you have the same Mainboard on your canvas as you actually have. Next drag on a Display, a Camera, and an SD Card, and arrange them as shown in Figure 2-2. Note that you do not have to add the red USB DP Power module to the Visual Studio Design. This will always be connected to connector 1.

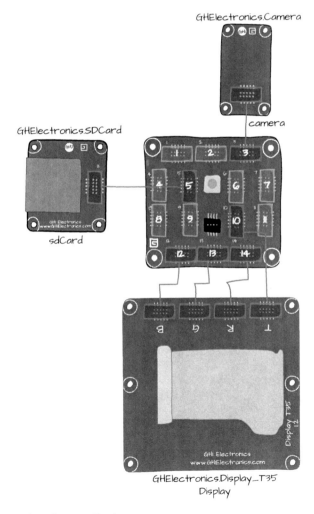

Figure 2-2. *Spy Camera Design*

Let's start by looking at the camera module (see Figure 2-3). This module is the kind of camera that is usually found in a webcam and therefore you are not going to be making a 14 megapixel camera out of it. The camera is manual focus and you will find that the front of the lens screws in and out to allow you to set the focus.

Figure 2-3. *Fez Spider Camera Module*

The camera's accompanying C# class allows it to operate in two modes. It can stream images in movie mode at up to 20 frames per second, or you can just trigger it to take separate images.

We will start by creating a simple program that takes a photograph every second and shows it on the display.

Save the design file (*Program.gadgeteer*) and then switch to Program.cs. Go through the generated code and modify it so that it appears as follows:

```
using System;
using Microsoft.SPOT;
using Microsoft.SPOT.Presentation;
using Microsoft.SPOT.Presentation.Controls;
using Microsoft.SPOT.Presentation.Media;

using GT = Gadgeteer;
using GTM = Gadgeteer.Modules;
using Gadgeteer.Modules.GHIElectronics;

namespace SpyCameraTest
{
    public partial class Program
    {

        GT.Timer timer = new GT.Timer(1000);

        void ProgramStarted()
        {
```

```
                timer.Tick += new GT.Timer.TickEventHandler(timer_Tick);
                timer.Start();
                camera.PictureCaptured += new
    Camera.PictureCapturedEventHandler(camera_PictureCaptured);
            }

            void camera_PictureCaptured(Camera sender, GT.Picture picture)
            {
                display.SimpleGraphics.DisplayImage(picture, 0, 0);
            }

            void timer_Tick(GT.Timer timer)
            {
                if (camera.CameraReady)
                {
                    camera.TakePicture();
                }
            }
        }
    }
```

You will notice a few new things about this code, so let's go through them.

First of all, we need a way to trigger the camera to photograph every second. We do this by using a timer. The following line defines a timer that will do something every second:

```
GT.Timer timer = new GT.Timer(1000);
```

We specify what the timer is to do every second in the ProgramStarted method by adding a handler to it using the syntax "+=":

```
timer.Tick += new GT.Timer.TickEventHandler(timer_Tick);
```

As soon as you key in the "=" of "+=", you will be prompted to press the Tab key to insert the code automatically. Do this; it saves a lot of typing.

The second prompt to press the tab key from Visual Studio will then create the actual handler method stub. You then have to complete the method so that it appears as follows:

```
void timer_Tick(GT.Timer timer)
{
    if (camera.CameraReady)
    {
        camera.TakePicture();
    }
}
```

This handler method will, first of all, see if the camera module is ready to take a photograph using the property CameraReady. If it is, then it will take a photograph.

NOTE: This is just asking the camera to take the photograph. When the camera has actually taken the photograph, another handler will run.

To see how that happens we need to return to the **ProgramStarted** method:

```
void ProgramStarted()
{
    timer.Tick += new GT.Timer.TickEventHandler(timer_Tick);
    timer.Start();
    camera.PictureCaptured += new
Camera.PictureCapturedEventHandler(camera_PictureCaptured);
}
```

After the timer has had a handler added to it to take photographs every second, we need to tell the timer to start:

```
timer.Start();
```

We then add a handler to the camera's **PictureCaptured** event. Once again, we can use the tab feature to create the handler method stub for us:

```
void camera_PictureCaptured(Camera sender, GT.Picture picture)
{
    display.SimpleGraphics.DisplayImage(picture, 0, 0);
}
```

When the event occurs and the handler is run, it receives arguments of the camera used to take the photo and the resultant picture. We are really interested only in the picture.

We can then display the picture on the **display**.

NOTE: For convenience, in this example we use the **SimpleGraphics** option on display to display the image. This approach works, but **SimpleGraphics** could perhaps also be called **SlowGraphics**, and as we develop this project further we will switch to using a different mechanism to show things on the display.

Trying It Out

Connect the modules to the Mainboard as shown in Figure 2-2 and deploy the project to the Gadgeteer. You should see the screen update with a new image from the camera every second. Remember that the camera is manual focus, so you will have to adjust the lens on the front to get a crisp image.

WPFWindow

This example is fine, but we need to be able to display a little more than just the last image captured by the camera. We need to be able to display messages, such as what the interval is set to, and a message if there is an error writing to the SD card.

To accommodate this message area, we need to abandon our SimpleGraph ics interface to the display and use something called WPFWindow.

WPFWindow is a class that allows us to structure the screen into various different areas. In this case, the whole screen as one area to display the image, and a small area near the top of the screen, where sometimes a message will appear.

We are going to modify the "Spy Camera Test" project to display a little notification at the top of the screen whenever a photo is taken.

This program is now over 100 lines long, so you will probably want to open the "Spy Camera Test" project now.

Creating a UI Structure

The biggest change to the project is the way the display is used. This is all contained in a new method called SetupUI (UI for User Interface). This method would normally appear in the code just after ProgramStarted:

```
void SetupUI()
{
    mainWindow = display.WPFWindow;

    // setup the layout
    Canvas layout = new Canvas();
    Border background = new Border();
    background.Background = new SolidColorBrush(Colors.Black);
    background.Height = 240;
    background.Width = 320;
    layout.Children.Add(background);
    Canvas.SetLeft(background, 0);
    Canvas.SetTop(background, 0);
```

```
// add the image display
imageDisplay = new Border();
imageDisplay.Height = 240;
imageDisplay.Width = 320;
layout.Children.Add(imageDisplay);
Canvas.SetLeft(imageDisplay, 0);
Canvas.SetTop(imageDisplay, 0);

// add the text label
label = new Text("");
label.Height = 50;
label.Width = 320;
label.ForeColor = Colors.White;
label.Font = Resources.GetFont(Resources.FontResources.NinaB);
label.TextAlignment = TextAlignment.Center;
label.Visibility = Visibility.Collapsed;
layout.Children.Add(label);
Canvas.SetLeft(label, 0);
Canvas.SetTop(label, 0);

mainWindow.Child = layout;
}
```

Figure 2-4 shows the structure of objects that SetupUI creates.

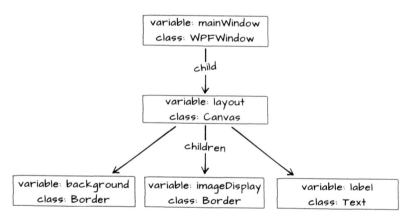

Figure 2-4. *Screen Objects*

The Canvas class is used as a container of other controls that are to appear on the screen. This will include two objects of type Border (background and imageDisplay) and an object of type Text (label).

The first section of code in SetupUI assigns a member variable the value of the display's WPFWindow attribute. The next section creates the layout canvas and also a background Border. Border is a bit of a misnomer, as it will in fact fill the whole screen and be a big black rectangle. This will be visible as a background to the label when the **imageDisplay** is hidden in order to display a message.

Next, the **imageDisplay** is created. This is much the same as the background, except that it will eventually contain an image from the camera.

Now we can add the label. This is an instance of the class **Text**. It occupies only the first 50 pixels in height of the screen.

 NOTE: Screen coordinates have an origin in the top-left corner of the screen and the y axis increases in value as you move down the screen. The x axis is left to right as you would expect. The screen we are using has a resolution of 320 x 240 pixels.

When the label is created, it is not shown initially, because of the following line:

```
label.Visibility = Visibility.Collapsed;
```

The final line in **SetupUI** sets the layout to be the child of the **mainWindow**. This means that whenever the **mainWindow** displays it will display its child. That child being the layout, will in turn display all its children. That is, the background, the imageDisplay, and the label.

Showing Messages

The mechanism for displaying a message relies on hiding the **imageDisplay** (making it invisible) and showing the label. However, we also want the message to automatically disappear after a second. We do this using another timer.

The method **DisplayMessage** takes a single string as its argument and will display it. This method can be positioned anywhere in the class at the same level as the other methods:

```
private void DisplayMessage(string message)
{
    // hide image
    imageDisplay.Visibility = Visibility.Collapsed;

    // show label
    label.Visibility = Visibility.Visible;
    label.TextContent = message;
    label.UpdateLayout();
```

```
    if (hideMessageTimer == null)
    {
        hideMessageTimer = new GT.Timer(1000);
        hideMessageTimer.Tick += new
GT.Timer.TickEventHandler(hideMessage_Tick);
        hideMessageTimer.Start();
    }
    else
    {
        hideMessageTimer.Restart();
    }
}
```

The first thing it does is hide the `imageDisplay`, so that we are just left with the black background visible. It then makes the label visible and sets its contents to the message that we want to display.

The rest of the method is concerned with creating a timer that will (after a second) hide the label again. If this is the first time the timer has been used, then it will not have been created, so the first time we call `DisplayMessage` `hideMessageTimer` will be `null`. However, on future calls to `DisplayMessage`, we can just call `Restart` on the timer.

The handler for the `Timer Tick` is straightforward:

```
void hideMessage_Tick(GT.Timer timer)
{
    timer.Stop();
    label.Visibility = Visibility.Collapsed;
    imageDisplay.Visibility = Visibility.Visible;
}
```

It stops the timer, then hides the label and makes the `imageDisplay` visible again.

The rest of the program is the same as our previous version.

The Final Program

It is now time to open the final project (*SpyCamera*) in Visual Studio and look at the changes that we need to make the final project. That is, to allow us to detect a touch on the screen, cycle through our picture capture intervals, and write the pictures to the SD card.

Touch Screen

The display we have used also has a touch screen. This can detect where someone has touched it. We are going to use a touch on the screen to change the time interval. Each time the screen is tapped, it will step along to the next time interval from an array of time intervals. Finally, when it gets to the end of the list it will change to a setting of "Camera Off."

Detecting a touch on the screen is as simple as adding this line to SetupUI:

```
mainWindow.TouchDown += new
Microsoft.SPOT.Input.TouchEventHandler(mainWindow_TouchDown);
```

Now, whenever the screen is touched, `mainWindow_TouchDown` will be called.

To be able to cycle through a series of intervals, we use the following data structure:

```
int[] intervals = {0, 5, 10, 20, 30, 60};
```

This creates an array of interval values in seconds. We then use another variable (`intervalPosition`) to identify the current interval. Then, when we want to move onto the next interval, we increment `intervalPosition` until we get to the end of the list and then go back to 0. An interval of 0 will have the special meaning of "Camera Off."

We can see this in the `mainWindow_TouchDown` handler:

```
void mainWindow_TouchDown(object sender,
Microsoft.SPOT.Input.TouchEventArgs e)
{
    intervalPosition++;
    if (intervalPosition == intervals.Length)
    {
        intervalPosition = 0;
    }
    setTimerInterval(intervalPosition);
}
```

Changing the interval of the timer itself occurs in the method `setTimerInterval`:

```
void setTimerInterval(int newIntervalPosition)
{
  intervalPosition = newIntervalPosition;
  TimeSpan interval = new TimeSpan(0, 0, 0, 0, intervals[intervalPosition]
* 1000);
  if (timer != null)
  {
    timer.Stop();
  }
  else
  {
    timer = new GT.Timer(interval);
    timer.Tick += new GT.Timer.TickEventHandler(timer_Tick);
  }
  if (intervalPosition == 0)
```

```
  {
    DisplayMessage("Camera Off");
  }
  else
  {
    timer.Interval = interval;
    timer.Start();
    DisplayMessage("Interval Set to: " + intervals[intervalPosition]
+ " seconds");
  }
}
```

Just like our messageTimer, we start by creating a new Timer if it hasn't yet
been created.

We then have a test to see if the intervalPosition is 0. If it is, then we display
the message "Camera Off," otherwise, we create a new Timespan object of
the duration indicated by our intervals array. This is then used to change the
interval of the Timer before restarting it.

Writing to SD cards

The final part of the project is writing the captured images to the SD card.
You will notice that the camera_PictureCaptured method has a new line that
assigns the image just taken to a member variable called lastPicture:

```
void camera_PictureCaptured(Camera sender, GT.Picture picture)
{
    imageDisplay.Background = new ImageBrush(picture.MakeBitmap());
    lastPicture = picture;
}
```

Now, whenever the main timer "ticks," we will save the previous image before
taking a new one:

```
void timer_Tick(GT.Timer timer)
{
    // save the last image captured.
    // The new one probably isn't ready yet
    SaveLastImage();
    if (camera.CameraReady)
    {
        camera.TakePicture();
    }
}
```

The actual writing to the SD card is all contained in the method SaveLast
Image:

```
void SaveLastImage()
{
```

```
    if (lastPicture == null)
    {
        return;
    }
    try
    {
        String filename = "picture_" + pictureIndex + ".bmp";
        DisplayMessage("Saving .....");
        sdCard.GetStorageDevice().WriteFile(filename,
lastPicture.PictureData);
        DisplayMessage("Photo Saved to: " + filename);
        pictureIndex++;
    }
    catch (Exception ex)
    {
        DisplayMessage("SD Card Error");
    }
}
```

First of all, this checks to see that there is an image to be saved. If there isn't, it simply returns without doing anything.

If there is an image, then the remainder of the method is contained in a try / catch block. This ensures that if anything goes wrong in saving the image—for example, the SD card is not inserted or is full or write-protected—then a message will be displayed.

The first thing we do is construct a filename for the image. We do this by concatenating the word "picture_" with a serial number and then adding ".bmp" to the end. The serial number is held in a variable pictureIndex that is incremented each time an image is saved.

Actually writing the data to the file is just a matter of calling WriteFile on the storage device with the binary data of the last image.

Testing

Deploy the project to your hardware and you should find that you can change the interval by touching the screen; if an SD card is inserted, files will be created.

Transferring the SD card to a computer, you should find that you can open and view the images from the camera.

Conclusion

This is quite a complicated first project, but we have explored the use of the display, as well as getting to grips with the camera and using SD cards.

In Chapter 3, we will develop a simple game that uses the display in a similar way to this project, but also introduces the joystick module.

3/Snowflakes Game

This is a simple game in which you guide a tongue back and forth across the screen in order to catch snowflakes that are falling from the sky.

The project uses the display and joystick modules included in the Fez Starter Kit.

The Design

Table 3-1. *You will need*

Item	Source
Project	*http://www.gadgeteerbook.com/downloads/3.1.Snowflake.zip*
Fez Spider Mainboard	*http://www.ghielectronics.com/catalog/product/269*
USB Client DP module	*http://www.ghielectronics.com/catalog/product/280*
Display T35 Module	*http://www.ghielectronics.com/catalog/product/276*
Joystick Module	*http://www.ghielectronics.com/catalog/product/299*

Once you have downloaded the zip files for the projects from the book's website (*http://www.gadgeteerbook.com/downloads*) you will need to unzip this into your Gadgeteer's projects area which is usually *My Documents \Visual Studio 2010\Projects*. You can then open the project by opening the File menu and selecting "Open Project.." from Visual Studio's menu.

The arrangement of the modules in this project is show in Figure 3-2.

Screen Layout

This project uses a similar window setup to the project of Chapter 2. This requires the member variables `mainWindow`, `canvas`, and `label`. It also defines two `Rectangles` (`tongue` and `snowflake`) that will also be part of the layout.

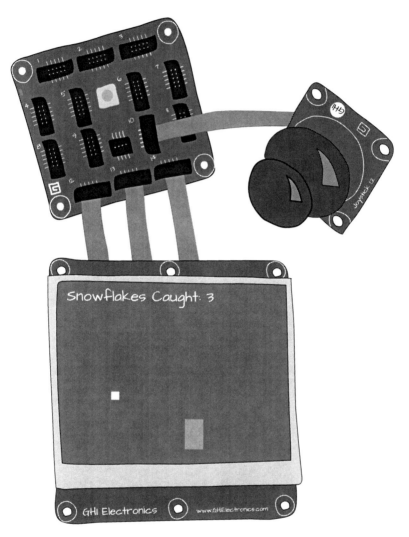

Figure 3-1. *Snowflake Game*

The animation effect of the game will be achieved by moving the positions of these rectangles.

You will also see a group of *int* variables:

```
int tongueLeftPosition = 150;
int snowflakeLeftPosition = 150;
int snowflakeTopPosition = 50;
int tongueWidth = 30;
```

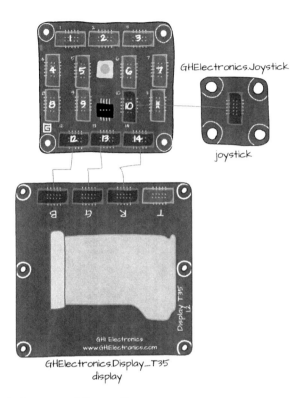

GHElectronics.Joystick

joystick

GHElectronics.Display_T35
display

Figure 3-2. *Component Connections*

These are used to manage the position of those rectangles in the game, as well as to define the size of the tongue and snowflake.

The project uses two timers: one to control the tongue, which is called every 30 milliseconds, and another that is responsible for the movement of the snowflake.

The apparently random movement of the snowflake uses a random number generator:

```
Random randomNumberGenerator = new Random();
```

One instance of this class is created, and every time we need a new random number, we can ask for one.

NOTE: These numbers are not truly random, they are actually in a sequence and are an example of pseudorandom numbers. If you are interested, you can find more information on pseudorandom number generation here: *http://en.wikipedia.org/wiki/Pseudorandom_number_generator*.

Figure 3-3 shows the structure of the screen. You can see how the back ground, label, and two Rectangles make up the items to be displayed.

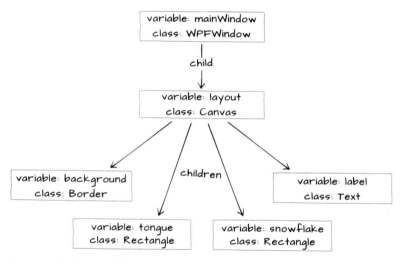

Figure 3-3. *Screen Objects*

The SetupUI method, which creates this structure, is quite similar to that of the previous project:

```
void SetupUI()
{
    // initialize window
    mainWindow = display.WPFWindow;

    // setup the layout
    layout = new Canvas();
    Border background = new Border();
    background.Background = new SolidColorBrush(Colors.Black);
    background.Height = 240;
    background.Width = 320;

    layout.Children.Add(background);
```

```
Canvas.SetLeft(background, 0);
Canvas.SetTop(background, 0);

//add the tongue
tongue = new Rectangle(tongueWidth, 40);
tongue.Fill = new SolidColorBrush(Colors.Red);
layout.Children.Add(tongue);

//add the snowflake
snowflake = new Rectangle(10, 10);
snowflake.Fill = new SolidColorBrush(Colors.White);
layout.Children.Add(snowflake);

// add the text area
label = new Text();
label.Height = 240;
label.Width = 320;
label.ForeColor = Colors.White;
label.Font = Resources.GetFont(Resources.FontResources.NinaB);

layout.Children.Add(label);
Canvas.SetLeft(label, 0);
Canvas.SetTop(label, 0);

mainWindow.Child = layout;
}
```

One difference between this and the previous project is that here, all parts of the display remain visible all of the time. There is no showing and hiding of controls.

SetupUI is called from the ProgramStarted method:

```
void ProgramStarted()
{
    SetupUI();

    Canvas.SetLeft(tongue, tongueLeftPosition);
    Canvas.SetTop(tongue, 200);

    Canvas.SetLeft(snowflake, snowflakeLeftPosition);
    Canvas.SetTop(snowflake, snowflakeTopPosition);

    joystickTimer.Tick += new GT.Timer.TickEventHandler(JoystickTimer_Tick);
    joystickTimer.Start();

    snowFlakeTimer.Tick += new
GT.Timer.TickEventHandler(SnowflakeTimer_Tick);
    snowFlakeTimer.Start();
}
```

The ProgramStarted then sets the positions of the *tongue* and *snowflake* rectangles on the layout Canvas using the methods SetLeft and SetTop. Later, we will use these same methods to change the positions of the tongue and snowflake. They will automatically handle any redrawing required, so we do not need to worry about erasing them from their old position before allowing them to be redrawn at their new one.

Finally, ProgramStarted creates the two timers and starts them.

Using the Joystick module

The joystick (shown in Figure 3-4)is a two-axis joystick combined with a switch, which is activated when the joystick button itself is pressed. We will not be using the joystick's button; we will just use the readings of the joystick on the x axis.

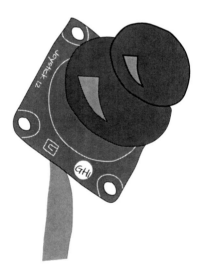

Figure 3-4. *Joystick*

At any time, your program can ask for the position of the joystick's position by calling GetJoystickPostion on the joystick variable. This call returns an instance of Joystick.Position. As in the case of this game, we are only interested in the "x" position of the joystick, we use the following:

```
double x = joystick.GetJoystickPostion().X;
```

This returns a number between 0.0 and 1.0, where 0.0 is far left, 1.0 far right and, yes, you guessed it, 0.5 is in the middle.

In this game, we will not use the absolute position of the joystick to set the position of the tongue but rather determine if the joystick is towards the left, towards the right, or in the middle.

The method to do this is as follows:

```
void JoystickTimer_Tick(GT.Timer timer)
{
    double x = joystick.GetJoystickPostion().X;
    if (x < 0.3 && tongueLeftPosition > 0)
    {
        tongueLeftPosition -= 5;
    }
    else if (x > 0.7 && tongueLeftPosition < 320 - tongueWidth)
    {
        tongueLeftPosition += 5;
    }
    Canvas.SetLeft(tongue, tongueLeftPosition);
    CheckForLanding();
}
```

We just test to see if the x position is less than 0.3 and if it is, assume that we want to move left. Similarly, if it is greater than 0.7, we move right. Otherwise, the tongue stays where it is.

You will notice the function call to CheckForLanding in JoystickTimer_Tick. This method will check for a collision between the snowflake and the tongue. If it finds one, it will add one to the score:

```
void CheckForLanding()
{
    if (snowflakeTopPosition > 200
        &&
        snowflakeLeftPosition + 10 >= tongueLeftPosition
        &&
        snowflakeLeftPosition <= tongueLeftPosition + tongueWidth)
    {
        score++;
        label.TextContent = "Snowflakes Caught: " + score;
        ResetSnowflake();
    }
}
```

It will help to refer to Figure 3-5 in working out what is going on with all the clauses "anded" together in the "if" statement.

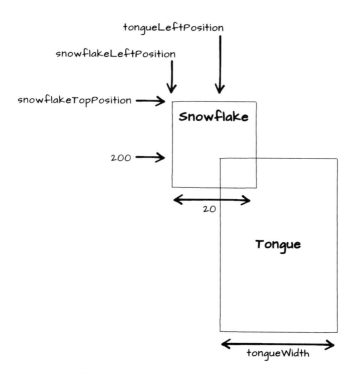

Figure 3-5. *Colliding Rectangles*

As we discussed earlier, the snowflake has its own timer, and each time the timer ticks, this method is run:

```
void SnowflakeTimer_Tick(GT.Timer timer)
{
    snowflakeTopPosition += 5;
    if (snowflakeTopPosition >= 240)
    {
        ResetSnowflake();
    }
    snowflakeLeftPosition += (randomNumberGenerator.Next(15) - 7);
    if (snowflakeLeftPosition < 10) snowflakeLeftPosition = 0;
    if (snowflakeLeftPosition > 300) snowflakeLeftPosition = 300;
    Canvas.SetLeft(snowflake, snowflakeLeftPosition);
    Canvas.SetTop(snowflake, snowflakeTopPosition);
}
```

The first thing we do is move the snowflake down the screen 5 pixels. We then check to see if the snowflake has reached the bottom of the screen, and if it has, we call the method **ResetSnowflake** that puts it back up at the top of the screen with a random position for its x coordinate.

The apparent random behavior of the snowflake is accomplished by using our random number generator to create an offset to snowflakeLeftPosition:

```
snowflakeLeftPosition += (randomNumberGenerator.Next(15) - 7);
```

The method Next returns the next random number from the random number generator. The argument specified a maximum value for the number, so in this case, it will be a number between 0 and 15.

If we just added this number to the snowflakeLeftPosition, the snowflake would only move to the right each time. To make sure that it can move to the left or right, we subtract 7 from our number between 0 and 15. So, for example if the random number was 7, the horizontal position would be unchanged, whereas if it was less than 7, the snowflake would move to the left by between 1 and 7 pixels. If the random number is greater than 7, the snowflake will move to the right.

Conclusion

This project is a very simple game, but you can probably see how you could modify it. You could for instance try adding a second snowflake, or even creating an array of snowflakes.

In the Chapter 4, we will go on to see how you can use .NET Gadgeteer as a mini web server running on your home network.

4/Web Messenger

The Gadgeteer with an Ethernet module can be put to work as a small web server. In this project, the web server constantly serves the contents of the Gadgeteer screen. You can draw sketches on the screen with your finger or a stylus and leave messages for the world to see, as shown in Figure 4-1.

Table 4-1. *You will need*

Item	Source
Simple Web Server Project	*http://www.gadgeteerbook.com/downloads/4.1.HelloWebServer.zip*
Sketch Pad Project	*http://www.gadgeteerbook.com/downloads/4.2.SketchPad.zip*
Full Project	*http://www.gadgeteerbook.com/downloads/4.3.WebMessenger.zip*
Ethernet patch cable	*http://www.sparkfun.com/products/8915*
Fez Spider Mainboard	*http://www.ghielectronics.com/catalog/product/269*
USB Client DP module	*http://www.ghielectronics.com/catalog/product/280*
Display T35 Module	*http://www.ghielectronics.com/catalog/product/276*
Multicolor LED	*http://www.ghielectronics.com/catalog/product/272*
Ethernet J11D Module	*http://www.ghielectronics.com/catalog/product/284*

All components, apart from the network patch cable are all included in the Fez Starter Kit. You can get the patch cable from any computer store.

Once you have downloaded the zip files for the projects from the book's website (*http://www.gadgeteerbook.com/downloads*) you will need to unzip them into your Gadgeteer's projects area which is usually *My Documents \Visual Studio 2010\Projects*. You can then open the project by opening the File menu and selecting "Open Project.." from Visual Studio's menu.

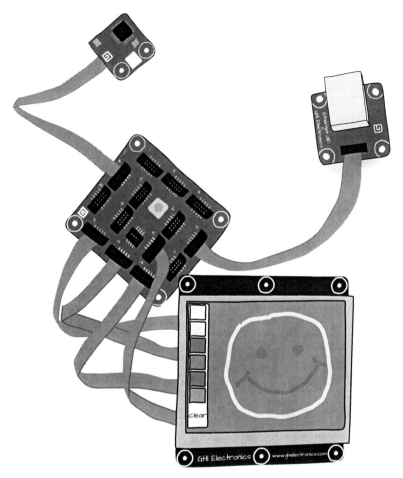

Figure 4-1. *Web Messenger*

Web Servers

Before we look at making our own web server, it is perhaps worth recapping exactly how a web server works.

A web server is a computer whose job it is to wait for a browser to request a URL from it and then to respond to it (see Figure 4-2).

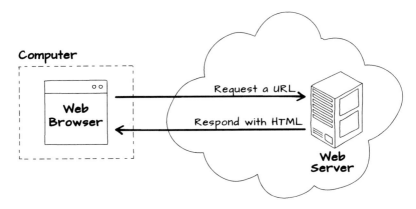

Figure 4-2. *Web Servers*

The response will be in the form of a stream of text formatted as HyperText Markup Language (HTML). HTML contains both text to be displayed and formatting that controls its appearance in the browser.

A Gadgeteer Web Server

The phrase "web page" is often used to describe a location on the Internet; you can think of a website as a whole load of web pages. However, you have seen that what is displayed on a particular "page" can be different every time you visit it. That is because the server creates the HTML before it eventually sends it back to the browser to be displayed.

In other words, web pages are often not just files of HTML that are sent back, but are constructed as a request arrives. The Gadgeteer approach to web serving is to construct the HTML it serves in response to a web request coming from the browser.

To learn a little about using the Ethernet module, we are going to start with a much simpler project, involving just the Ethernet module and a multicolor LED. We are going to make a minimal web server that will display a page that says "Hello World!!" (see Figure 4-3).

Figure 4-4 shows the modules from the Visual Studio designer.

When making a web server, we need to connect our Gadgeteer to the Internet, or at the very least to our local network. If you have Broadband Internet, then you are likely to have a modem/switch that may provide you with wireless access to the Internet, but more importantly for this project, a set of RJ45 sockets on the back (often four of them) to which we can connect the Gadgeteer using an Ethernet patch lead.

Figure 4-3. *Hello World*

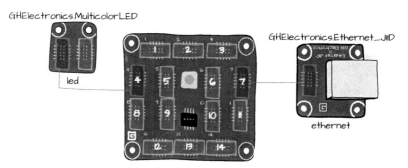

Figure 4-4. *Hello Web Server Modules*

While we are programming it, the Gadgeteer will also be connected to our computer, which will probably also be providing it power from the USB connection. Once the Gadgeteer is programmed we don't need the USB connection anymore, and could power the Gadgeteer from an external power supply. All these connections are summarized in Figure 4-5.

The Gadgeteer Mainboard will be assigned an IP address on the local network. You will then be able to visit a page on the Gadgeteer from a browser on any computer or mobile Internet device connected to the same network.

It is actually remarkably easy to turn a Gadgeteer into a web server. So, without further ado, let's have a look at some of the code.

We have already seen how useful it is to attach event handlers to objects. For instance, to attach a listener to a button, so that when it is pressed a

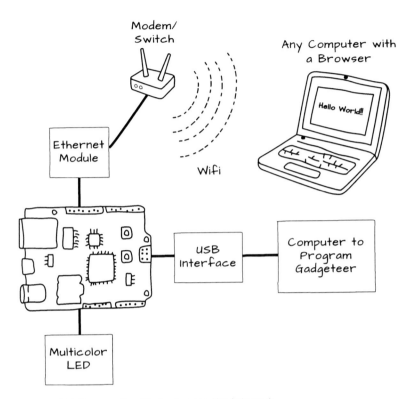

Figure 4-5. *Connecting Gadgeteer to the Internet*

method is run. The .NET Gadgeteer approach to web serving also makes good use of event handlers:

```
namespace HelloWebServer
{
  public partial class Program
  {
    GT.Networking.WebEvent sayHello;

    void ProgramStarted()
    {
      ethernet.UseDHCP();
      ethernet.NetworkUp += new
GTM.Module.NetworkModule.NetworkEventHandler(ethernet_NetworkUp);
      ethernet.NetworkDown += new
GTM.Module.NetworkModule.NetworkEventHandler(ethernet_NetworkDown);
      led.TurnBlue();
    }

        void ethernet_NetworkUp(GTM.Module.NetworkModule sender,
```

```
GTM.Module.NetworkModule.NetworkState state)
    {
      led.TurnGreen();
      string ipAddress = ethernet.NetworkSettings.IPAddress;
      WebServer.StartLocalServer(ipAddress, 80);
      sayHello = WebServer.SetupWebEvent("hello");
      sayHello.WebEventReceived += new
WebEvent.ReceivedWebEventHandler(sayHello_WebEventReceived);
    }

        void sayHello_WebEventReceived(string path, WebServer.HttpMethod
method, Responder responder)
    {
      string content = "<html><body><h1>Hello World!!</h1></body></html>";
      byte[] bytes = new System.Text.UTF8Encoding().GetBytes(content);
      responder.Respond(bytes, "text/html");
    }

        void ethernet_NetworkDown(GTM.Module.NetworkModule sender,
GTM.Module.NetworkModule.NetworkState state)
    {
      led.TurnRed();
    }
  }
}
```

The ProgramStarted method first tells the ethernet module to use DHCP. DHCP is a mechanism that automatically assigns an IP address to a device when it is connected to the network. If you have a home network, chances are it will be set up for DHCP.

A device's Internet Protocol (IP) address uniquely identifies the device on the network, which is why it is important that the Gadgeteer does not end up with an IP address that is already in use.

IP addresses are strange looking numbers that come in four parts. A typical IP address for use inside the network (the case here) might look something like: 192.168.1.106 and as you can see from Figure 4-3 you can use this in a URL instead of a domain name.

ProgramStarted also attaches handlers to the events NetworkUp and NetworkDown.

The handler for NetworkUp sets the LED to green and then starts a web server running determining the IP address of the Gadgeteer.

TIP: In some early LED modules, the green and blue LEDs are swapped over, and so the wrong color is displayed. If you have such an LED, then you can fix it by adding this line to the start of the `ProgramStar ted` method:

```
led.GreenBlueSwapped = true;
```

To understand the rest of what `ethernet_NetworkUp` does, we need to step back a little and look at the top of the program where we define `sayHello`:

```
GT.Networking.WebEvent sayHello;
```

The event `sayHello` is a `WebEvent` and to some extent, it is similar to the idea of a web page. In this example, we always display the same thing in the browser, but in a more complex example, we might set up several `WebE vents` for different actions on the web server.

The method `ethernet_NetworkUp` makes the web server aware of the `say Hello WebEvent` using the following line:

```
sayHello = WebServer.SetupWebEvent("hello");
```

The string "hello" associates the name "hello" with this web event. That means that if we put "/hello" on the end of our URL in the browser, it is this `WebEvent` that will be invoked. That is why we attach yet another handler, this time to the event `WebEventReceived` of `sayHello`.

The handler method needs to construct some HTML to be displayed on the browser. The HTML that we are going to return is:

```
<html><body><h1>Hello World!!</h1></body></html>
```

The bits between "<" and ">" are called tags and should come in matching beginning and ending tags. Notice how the end tags have a "/" right after the "<".

A web page should always contain a <body> tag within an <html> tag. That just leaves the text "Hello World" inside the <h1> tag. The tag "h1" is used to indicate a level 1 heading, which is why our browser displays it in a large font (see Figure 4-3).

There is much more to HTML than the few tags we have looked at here. Search the Internet for "HTML Tutorial" for more information on writing HTML.

The sayHello_WebEventReceived event handler is supplied with an argument of responder. It is this that we must use to send the HTML back to the browser. The method to do this is not unsurprisingly called Respond:

```
void sayHello_WebEventReceived(string path, WebServer.HttpMethod method,
    Responder responder)
{
    string content = "<html><body><h1>Hello World!!</h1></body></html>";
    byte[] bytes = new System.Text.UTF8Encoding().GetBytes(content);
    responder.Respond(bytes, "text/html");
}
```

However, Respond does not take a string, but rather a byte array, so we need to do a little magic to convert our string into a byte array:

```
byte[] bytes = new System.Text.UTF8Encoding().GetBytes(content);
```

The second argument to Respond tells the browser what kind of response to expect to be contained in the byte array, and "text/html" is the standard way of saying it is HTML. In other circumstances, for example, it could be an image. We will be sending images later, when we combine this with our finger painting code.

We haven't described the ethernet_NetworkDown event handler. All it does is set the LED color to red to show that we do not have a network connection.

Testing

We are now ready to test our little web server, so connect the modules according to Figure 4-4 and program the Mainboard with the HelloWebServer program. You can also plug an Ethernet cable between the Ethernet module and your modem/switch. Don't forget to also attach the DP Power module to connector 1.

After a short pause, the LED should turn green. Success!

Sometimes you have to unplug and the plug the Ethernet cable back in to make Gadgeteer aware of the connection. If this does not work and you get error messages when you deploy, then you might have an old version of the Gadgeteer Firmware and/or the SDK, so follow the installation and firmware update procedure described in "Installation" on page 5.

We do, however, have one little problem to overcome. Since the IP address is allocated automatically, we don't know what we need to type into our browser.

To discover this, we need to look at the output window of Visual Studio, as shown in Figure 4-6. This window is only visible when you are in the debugger. If it is not showing, then you may need to make it visible by dragging the

bottom partition of the screen—where the Error List usually resides—
up a little.

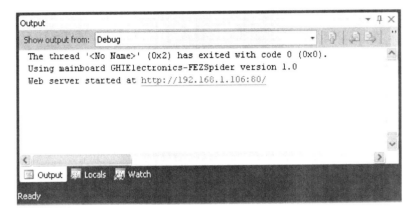

Figure 4-6. *IP Address in the Output Window*

You can see the following line in the output:

```
Web server started at http://192.168.1.106:80/
```

This tells us that the sever is up and running and listening for requests on
port 80 (the :80 part). Since port 80 is the standard port, it does not need
to be included in the URL.

Most home modem/switches will keep the same IP address for a device once
it has been allocated, so you should not need to do this every time. If you do
find that the IP address changes then go to the web console for your home
Broadband modem / switch and look for an option of the DHCP page to set
the lease time to as long as possible.

You can now open your browser and enter the URL for your IP address with
"/hello" on the end, and you should see something like Figure 4-3.

A Gadgeteer Sketch Pad

In this section, we are going to create a Gadgeteer Sketch Pad, before com-
bining it with the Ethernet project to allow our drawings and messages to be
served up over the network.

This project is available as a separate project without the Ethernet part and
it doesn't need the Ethernet module or multicolor LED. In fact, it is just a
Mainboard and Display.

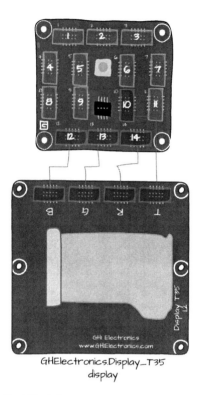

GHElectronicsDisplay_T35
display

Figure 4-7. *Sketch Pad Modules*

The SketchPad screen (see Figure 4-8) has two areas, a main drawing area on the right and a palette area on the left where you can select the drawing color, or click Clear to clear the drawing area. You can erase by selecting black as the drawing color.

In the discussion that follows, you should open the SketchPad project in Visual Studio so that you can see all the code.

This is another project where we use the WPFWindow class. As usual, the window has a Canvas to contain our components. In this case, the components are an Image named background, which is used as the drawing area, and a set of Border objects that are used as the buttons in the palette area:

```
void SetupUI()
{
    Border whiteButton;
    Border yellowButton;
    Border redButton;
    Border greenButton;
    Border blueButton;
```

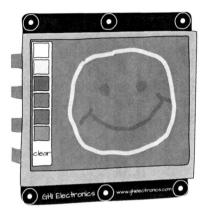

Figure 4-8. *Sketch Pad Screen*

```
Border blackButton;
Text clearButton;
// initialize window
mainWindow = display.WPFWindow;

// setup the layout
layout = new Canvas();
background = new Image();
background.Bitmap = new Bitmap(320 - sideBarWidth, 240);
background.Height = 240;
background.Width = 320 - sideBarWidth;

layout.Children.Add(background);
Canvas.SetLeft(background, sideBarWidth);
Canvas.SetTop(background, 0);

whiteButton = new Border();
SetupButton(whiteButton, Colors.White, 0);
whiteButton.TouchUp += new
Microsoft.SPOT.Input.TouchEventHandler(whiteButton_TouchUp);

          ......... other buttons not shown for clarity

          clearButton = SetupClearButton();

mainWindow.Child = layout;
```

As you can see, the color buttons are set up using a separate method, since
the initialization code is very similar for each button:

```
private void SetupButton(Border button, Color color, int position)
{
    button.Height = buttonHeight;
```

```
button.Width = buttonWidth;
button.Background = new SolidColorBrush(color);
layout.Children.Add(button);
Canvas.SetLeft(button, 2);
Canvas.SetTop(button, 2 + position * (buttonHeight + 2));
}
```

Each of the color buttons has its own handler method that simply changes the color of a member variable `drawingColor`:

```
void redButton_TouchUp(object sender, Microsoft.SPOT.Input.TouchEventArgs e)
{
    drawingColor = GT.Color.Red;
}
```

The actual drawing takes place in the following method:

```
void mainWindow_TouchMove(object sender,
Microsoft.SPOT.Input.TouchEventArgs e)
{
    TouchInput[] touches = e.Touches;
    int oldX = -1;
    int oldY = -1;
    foreach (TouchInput touch in touches)
    {
        int x = touch.X - sideBarWidth;
        int y = touch.Y;
        if (oldX != -1)
        {
            background.Bitmap.DrawLine(drawingColor, 3, oldX, oldY, x, y);
        }
        oldX = x;
        oldY = y;
    }
    background.Invalidate();
}
```

The method `mainWindow_TouchMove` is called every time you drag your finger around the screen. The result of one of these events is an array of touch positions that are found using:

```
TouchInput[] touches = e.Touches;
```

We can display such a sequence of points by drawing a line from the previous touch position to the new position. We keep the previous position in the variables `oldX` and `oldY`:

```
if (oldX != -1)
{
    background.Bitmap.DrawLine(drawingColor, 3, oldX, oldY, x, y);
}
oldX = x;
oldY = y;
```

The if statement ensures that we don't get a line from whatever oldX and oldY start at to the first touch position. Having drawn the line in the drawing Color, we must update oldX and oldY to be the current position ready for next time round the loop.

The following line of code ensures that the screen is refreshed to show the drawing:

```
background.Invalidate();
```

Putting It All Together

We now have everything we need for the project, we just need to combine the "HelloWebServer" and "SketchPad" projects into one, and add some code to serve back the image on the screen rather than just text.

Load the project "WebMessenger" into Visual Studio.

Figure 4-10 shows the arrangements of the modules and as you would expect, we use a combination of all the modules from the "HelloWebServer" and "SketchPad" projects.

One of the first things to notice in the code is that there is a new "using" line at the top. This is to include a utility method that we need to use to convert the bitmap on the screen to a BMP file format:

```
using GHIElectronics.NETMF.System;
```

This method is to be found in an assembly that is not automatically included in the project. So, if you were starting your own project that needed to use this, you would have to refer to it from your project.

To do this, right-click over the project, select the option "Add Reference...", then select "GHIElectronics.NETMF.System" from the list (see Figure 4-9).

Documentation for this method and other useful things can be found here: *http://www.ghielectronics.com/downloads/NETMF/Library%20Documentation/*

We define a WebEvent and call it sketch. The following shows the handler for WebEventReceived for this WebEvent:

```
void sketch_WebEventReceived(string path, WebServer.HttpMethod method,
Responder responder)
{
  Bitmap bitmap = background.Bitmap;
  byte[] buff = new byte[bitmap.Width * bitmap.Height * 3 + 54];
  Util.BitmapToBMPFile(bitmap.GetBitmap(), bitmap.Width, bitmap.Height,
buff);
  GT.Picture picture = new GT.Picture(buff, GT.Picture.PictureEncoding.BMP);
```

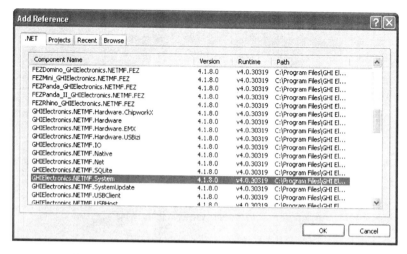

Figure 4-9. *Referencing a Package*

```
        responder.Respond(picture);
    }
```

The method `BitmapToBMPFile` requires a byte array buffer into which it puts the data for the BMP image. The size of this buffer must be the width of the image times its height x 3 + 54. The "times 3" is because each pixel is made up of 3 bytes, one each for the red, green, and blue components of the pixel.

The buffer length has 54 added to it to allow room for the BMP file's header.

A new `Picture` is created from the buffer and used as the argument to `Respond`.

Testing

Connect together the modules and load the project onto the Mainboard. The sketch controls and an empty screen should appear. The LED will be red, so plug the Ethernet cable in and it should turn green. As with the "HelloWebServer" example with which we started this chapter, make a note of the Mainboard's IP address in the Output are of Visual Studio.

Start a browser on a computer on the same network as the Mainboard and navigate to your IP address with "/sketch" appended. You should see an empty black image. This is to be expected, as we haven't drawn anything yet, so write a message or draw something on the screen and then hit Refresh.

You should see your message on the browser, a bit like Figure 4-11.

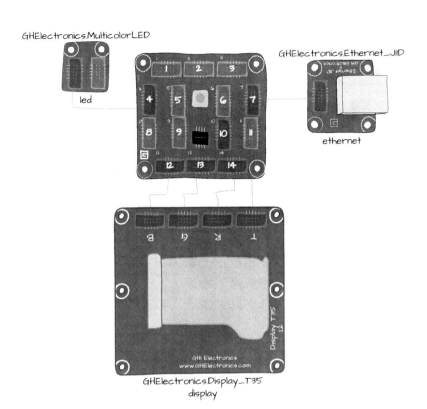

GHElectronics.MulticolorLED

led

GHElectronics.Ethernet_J11D

ethernet

GHElectronics.Display_T35
display

Figure 4-10. *Web Messenger Modules*

Sharing with the World

This is all very well and will allow you to share your messages inside your own network, but what if you want to share them with the world? This is accomplished by opening up a connection through your home modem/switch, so that the Gadgeteer mainboard is available as a web server to the rest of the world.

 WARNING: This has security implications for your home network, so do this only if you know what you are doing and are willing to take the risk.

There are various ways of doing this, but the safest is probably to go to your home network's administration console and on the Virtual Server section of the administration console, look for a port forwarding option. Use it to enable

Figure 4-11. *A Picture Message*

port forwarding on port 80 for the Gadgeteer Mainboard's internal IP address (something like 192.168.1.112).

To access the sketch from outside, you will need to use the external IP address of your home network. You will usually find this on the main status page of your administration console. But beware, because unless your Internet Service Provider provides a static IP address, this may change every time you restart your router.

Enter the external IP address with "/sketch" on the end and you should see the same image as you do from inside the network.

Conclusion

This is an interesting project and one that could be adapted for other purposes, perhaps with two Gadgeteers communicating a shared screen, or by making use of the camera module.

In Chapter 5, we attempt to get even more practical and create an SD card backup gadget.

5/Camera Backup Gadget

This project (shown in Figure 5-1) is intended for the digital camera user who, while travelling, might feel insecure about having all of his or her precious photographs on just one SD card that could become damaged or lost. Instead of copying the files to a laptop, this gadget will copy everything on an SD card to a USB flash drive.

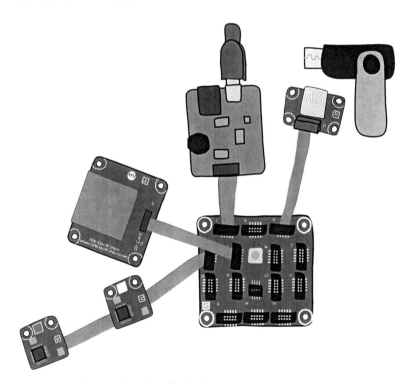

Figure 5-1. *Camera Backup Gadget*

The Design

Table 5-1. *What you will need*

Item	Source
Full Project	*http://www.gadgeteerbook.com/ downloads/5.1.CameraBackup.zip*
Fez Spider Mainboard	*http://www.ghielectronics.com/cata log/product/269*
USB Client DP module	*http://www.ghielectronics.com/cata log/product/280*
USB Host Module	*http://www.ghielectronics.com/cata log/product/270*
2 x Multicolor LED Module	*http://www.ghielectronics.com/cata log/product/272*
SD Card Module	*http://www.ghielectronics.com/cata log/product/271*
SD Card	
USB Flash Drive	

The components (apart from the SD Card and USB Flash Drive) are all included in the Fez Starter Kit.

Once you have downloaded the zip files for the projects from the book's website (*http://www.gadgeteerbook.com/downloads*) you will need to unzip them into your Gadgeteer's projects area which is usually *My Documents \Visual Studio 2010\Projects*. You can then open the project by opening the File menu and selecting "Open Project.." from Visual Studio's menu.

The project uses two multicolor LED modules. One shows the status of the SD card and the other of the USB flash drive. So when there is an SD card inserted, that LED will change from red to green, and when the USB storage device is inserted into the USB host module, its LED will similarly go green. When both LEDs are green, the transfer process will start automatically and as all the image files are copied from the SD card, the LEDs will flash green, turning solid green when all the files have been copied.

If anything goes wrong during the copying process, such as the USB storage becoming full, the LEDs will flash red.

Figure 5-2 shows the arrangement of the modules. Notice how one of the LEDs is connected to the other rather than to the Mainboard. The LED modules have two connectors to allow you to daisy-chain a series of LEDs without using up all the connectors on the Mainboard.

You can also see that the LED modules have been renamed to give them the more meaningful names of "ledSD" and "ledUSB". You can change the names of any of the components in the Visual Studio designer by right-clicking on them and selecting "Rename..".

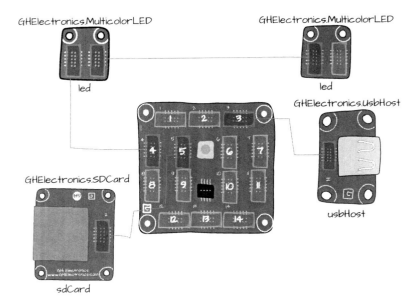

Figure 5-2. *Modules*

The USB Host Module

The USB Host module (shown in Figure 5-3) is quite a flexible module. In this project, it is used to connect a USB pen drive. However, it can also be used to attach a USB keyboard or mouse.

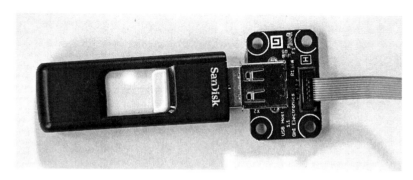

Figure 5-3. *USB Host Module*

The Program

To follow this discussion of the code, you should load the project into Visual Studio.

```
void ProgramStarted()
{
  usbHost.USBDriveConnected +=
   new UsbHost.USBDriveConnectedEventHandler(usbHost_USBDriveConnected);
  usbHost.USBDriveDisconnected +=
   new
UsbHost.USBDriveDisconnectedEventHandler(usbHost_USBDriveDisconnected);
  sdCard.SDCardMounted +=
   new SDCard.SDCardMountedEventHandler(sdCard_SDCardMounted);
  sdCard.SDCardUnmounted +=
   new SDCard.SDCardUnmountedEventHandler(sdCard_SDCardUnmounted);
  ledSD.TurnRed();
  ledUSB.TurnRed();
}
```

The flow of the program is totally controlled by the connected and disconnected event handlers for the SD card and USB host modules:

```
void sdCard_SDCardMounted(SDCard sender, GT.StorageDevice storageDevice)
{
    ledSD.TurnGreen();
    sdStorageDevice = storageDevice;
    CheckAndTransfer();
}
```

The disconnected handlers set the appropriate LED to red and set the member variables used to refer to the storage devices of the SD Card and USB module to null.

Each of the connected handlers calls the method `CheckAndTransfer` after setting the LED to green and assigning the `StorageDevice` for the SD card to a member variable `sdStorageDevice`. This allows the storage devices to be referenced by the `TransferFiles` method, whichever event handler is calling it. The two storage device variables also act as flags to indicate whether the storage device is ready to attempt the file transfer process. It they are null, they are not ready:

```
private void CheckAndTransfer()
{
  if (sdStorageDevice != null && usbStorageDevice != null)
  {
    TransferFiles();
  }
}
```

The `CheckAndTransfer` method tests to see if both the SD card and USB device are instantiated. If they are, it calls the `TransferFiles` method.

Copying Files

The `TransferFiles` method first sets both LEDs to blink green and then starts the actual copying process, which takes place in the `DeepCopy` method shown in the following code. After the deep copy has completed then both LEDs will be set to green again to indicate a successful copy.

The call to `DeepCopy` is contained within a **try catch** block. This ensures that if anything goes wrong inside the **try** block, the code in the **catch** block will be run. This will set both LEDs to flash red.

Try/Catch blocks are a useful mechanism for handling situations where things can go wrong. Not more so than in a situation where the program code is dependent on real hardware. For instance, unplugging the USB flash drive while copying is in progress:

```
private void TransferFiles()
{
    ledSD.BlinkRepeatedly(Colors.Green);
    ledUSB.BlinkRepeatedly(Colors.Green);
    try
    {
        CopyFiles("\\");
        DeepCopy(sdStorageDevice.ListRootDirectorySubdirectories());
        ledSD.TurnOff();
        ledSD.TurnGreen();
        ledUSB.TurnOff();
        ledUSB.TurnGreen();
    }
    catch (Exception)
```

```
        {
            ledSD.BlinkRepeatedly(Colors.Red);
            ledUSB.BlinkRepeatedly(Colors.Red);
        }
    }
```

One of the problems of copying the files from an SD card that is used in a camera is that the image files are always contained in a folder, or even in a folder inside another folder. So, to be sure of finding every image file on the SD card, we need to check every folder at the top level of the SD card, copy all of the files in that folder and if that folder contains a folder, check that folder and write any files there and any files contained in folders in that folder and so on.

You can see how there is a pattern here. What we are doing is known in computer science as a *depth-first recursive search*.

The DeepCopy method takes a list of directories as its argument. The first time this method is called from TransferFiles, the following will be the top-level directories:

```
sdStorageDevice.ListRootDirectorySubdirectories()
```

DeepCopy iterates over each of those directories, and as long as the directory does not begin with a "." the files in that directory will be copied and then the clever stuff happens.

The clever stuff is that DeepCopy then calls itself, but with an argument of the directories contained in the directory just processed.

Folders beginning with a "." are ignored, as the "." prefix is used in some operating systems to mean a hidden file. For example, on a Mac, an SD card will often contain a folder called ".Trashes" that contains any files that are put in the trash when the SD card is mounted on the Mac.

```
    private void DeepCopy(string[] sourceDirs)
    {
        foreach (string sourceDir in sourceDirs)
        {
            // ignore folders starting with .
            if (!sourceDir.Substring(0, 1).Equals("."))
            {
                CopyFiles(sourceDir);
                DeepCopy(sdStorageDevice.ListDirectories(sourceDir));
            }
        }
    }
```

This trick of a method calling itself is called *recursion*.

The `CopyFiles` method takes a directory as its argument and iterates over each file in the directory calling `BufferedFileCopy` on it.

 NOTE: As with directories, files whose names start with a "." are ignored.

```
private void CopyFiles(string sourceDir)
{
    string[] files = sdStorageDevice.ListFiles(sourceDir);
    foreach (string filepath in files)
    {
        if (!filepath.Substring(0, 1).Equals("."))
        {
            BufferedFileCopy(filepath);
        }
    }
}
```

`BufferedFileCopy` copies a file, but rather than read the entire contents of the file in one go, it reads it in blocks of 4096 bytes. Since image files can be large and video files even larger, this will prevent the Gadgeteer from running out of memory when copying.

Copying files in this way involves opening two `FileStreams`. The idea of `Streams` is that you can read or write data to or from them in chunks or as single bytes. In this case, we have an `outStream` opened on the new file that we are writing to on the USB device and an `inStream` on the file being copied on the SD card.

A byte array is used as a buffer to transfer up to 4096 bytes of data at a time.

When the method `Read` is called on the `inStream`, it places the data into the buffer and returns how much data was read. This will usually be 4096 bytes (a full buffer's worth) except when we are at the end of the file.

This is then written to the `outStream` and we repeat the whole process until there are no bytes left to read:

```
private void BufferedFileCopy(string filepath)
{
    int bufferSize = 4096;
    string[] parts = filepath.Split('\\');
    string filename = parts[parts.Length - 1];
    FileStream outStream = usbStorageDevice.OpenWrite(filename);
    FileStream inStream = sdStorageDevice.OpenRead(filepath);
    byte[] buffer = new byte[bufferSize];
    int bytesRead = inStream.Read(buffer, 0, bufferSize);
    while (bytesRead > 0)
```

```
    {
        outStream.Write(buffer, 0, bytesRead);
        bytesRead = inStream.Read(buffer, 0, bufferSize);
    }
    outStream.Close();
    inStream.Close();
}
```

When all the file has been copied, we close the streams.

Testing

This is just a matter of finding an SD card with some files on it and a USB drive with enough free space, then plugging them into the SD card module and USB sockets.

The LEDs should turn green and then flash while the files are being copied. This may take some time if the files are big. When the copying is complete, unplug the USB flash drive and plug it into your PC. You should find all the files on the SD card are now on it.

Breakpoints

In all the projects in this book, we have been using Visual Studio's debugger, while at the same time, not actually making use of its features.

The debugger is a very powerful tool for—well, you guessed it—debugging a project. It allows you to set breakpoints where the program will stop and you can inspect the values of variables, then move the program on one line at a time.

As an experiment, we are going to add a breakpoint to the program. To do this, all we have to do is click on the gray area to the left of the line of code where we want the program to pause (see Figure 5-4). In this case, on the second line of the CopyFiles method. Note how a big red dot has appeared next to the line. We can cancel the breakpoint by clicking on this dot.

Click the green play triangle to start the debugger as you normally would when deploying a project. Insert the USB storage device, but not the SD card yet.

It may seem like nothing is different, but actually, we will not get to the breakpoint until file copying begins when the SD card has been inserted, so insert the SD card to trigger the breakpoint.

You should now get to the breakpoint. Notice the change of appearance of the breakpoint dot, shown in Figure 5-5.

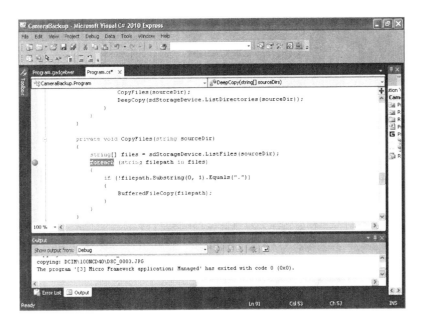

Figure 5-4. *Setting a Breakpoint*

If you click the Locals tab at the bottom of the screen, you can see that sourceDir has the value "\\" (the root directory) and files is an array of two folder names. You can also see the four control buttons that you can use to control the code in Figure 5-5.

"Continue" will resume the program until the next time that the breakpoint is hit.

"Step into" will jump into a method that is about to be called.

"Step Over" is the most commonly used command, this will step onto the next line. You should try clicking this a few times to get the idea.

"Step Out" will continue with the current method, stepping back out to the calling method.

You will often find that by putting the breakpoint in just the right place, you will not do much more than look at the variables and maybe "step over" a few times.

Debugging is a useful technique, so take some time to get comfortable with it; it will save you a lot of time in the long run.

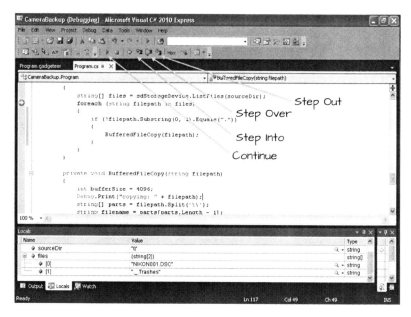

Figure 5-5. *Reaching a Breakpoint*

Conclusion

This is our final project of the book. It is one that could be boxed up and make a useful gadget.

In the final chapter of this book, we look at where to go next now that you have learned the basics of .NET Gadgeteer.

The pulse Oximeter measures variations in the amount of light passing through your finger as the blood pulses through.

Again, this uses event handlers, but this time the event is triggered every time a pulse is detected, rather than at a fixed period. This allows you to, for example, flash an LED each time a pulse is detected:

```
void ProgramStarted()
{
  pulseOximeter.Heartbeat +=
    new PulseOximeter.HeartbeatHandler(pulseOximeter_Heartbeat);
}

void pulseOximeter_Heartbeat(PulseOximeter sender,
                             PulseOximeter.Reading reading)
{
    lED7R.Animate(50, true, true, false);
    Debug.Print("Pulse Rate: " + reading.PulseRate);
}
```

Cellular Radio

This sophisticated module is essentially a cellular phone module. If you insert a SIM card, then you can use it to send data over the cellular network or send SMS text messages. It also has sound in and out, so could form the basis of your own cellphone.

Relay Module

Relays are simple electromechanical components that act as a high current switch. The Relay Module, shown in Figure 6-2, allows your Gadgeteer Mainboard to switch electrical items on and off.

Each of the four relays is capable of switching 15A at 120V, however, the printed circuit board on which the relays are mounted may not have tracks wide enough to support that, so check the manufacturer's specification first.

Convenient screw terminals are provided to which the devices you want to turn on and off are attached.

Music

The Music Module allows you to construct your own MP3 player. It can play music encoded as MP3, WMA, OGG, and WAV.

You will need to combine it with an SD Card or USB host module to act as storage for the music.

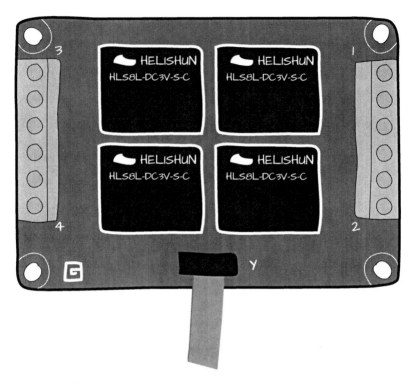

Figure 6-2. *Relay Module*

WiFi

In this book, we have made good use of the Ethernet module, but sometimes, it's nice not to have to be wired to the network. The WiFi module offers a wireless alternative to the Ethernet module.

Physical Design Files

If you follow one of the module URLs listed at the beginning of each chapter, go to the Downloads tab and you can download a 3D PDF file of the module. This is an accurate 3D model of the module, that you can use with 3D design software when designing an enclosure for your project.

NOTE: If your 3D design tool does not support 3D PDFs, Gadgeteer users have collected together 3D design files in other formats such as Google Sketchup's SKP format. They can be found here: *http://wiki.tinyclr.com/index.php?title=Gadget eer_3D_Models.*

Figure 6-3 shows one such 3D design for the T35 Display in Google Sketchup.

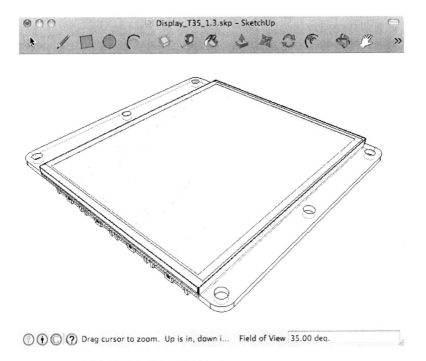

Figure 6-3. *A 3D Model of the T35 Display*

Documentation

A lot of the modules have the same style of programming interface, and you will often be able to work out how to use them by just typing the module's name and a "." then looking at the list of events and methods available for the module.

If you need to find out a bit more, the TinyCLR support page (*http://www.tinyclr.com/support*) offers a good starting point for documentation and tutorials.

If you are having trouble understanding how to use a module, have a look at the API Reference documents listed on that page. It can be a little difficult to navigate around, but basically this is where you will find a description of any method that you are trying to use.

Blogs and Web Resources

The .NET Gadgeteer community is full of enthusiastic Gadgeteers who are happy to help out. So if you find yourself stuck on something, a quick message on the TinyCLR forum (*www.tinyclr.com/forum/*) will quite often turn up someone who can point you in the right direction.

The TinyClr Wiki (*http://wiki.tinyclr.com/index.php?title=Main_Page*) is also a useful resource.

Conclusion

That concludes this book; it just remains for me to say that I hope you have enjoyed reading it and getting started with your Gadgeteer as much as I have enjoyed writing it.

The book's website (*http://www.gadgeteerbook.com*) will be kept up to date with errata and updates to the code.

About the Author

Dr. Simon Monk has a degree in Cybernetics and Computer Science and a PhD in Software Engineering. Simon spent several years as an academic before he returned to industry, co-founding the mobile software company Momote Ltd. He has been an active electronics hobbyist since his early teens. Simon is author of a number of hobby electronics books including *30 Arduino Projects for the Evil Genius*, *15 Dangerously Mad Projects for the Evil Genius*, and *Arduino + Android Projects for the Evil Genius*.

Have it your way.

Get even more for your money.

Join the O'Reilly Community, and register the O'Reilly books you own. It's free, and you'll get:

- $4.99 ebook upgrade offer
- 40% upgrade offer on O'Reilly print books
- Membership discounts on books and events
- Free lifetime updates to ebooks and videos
- Multiple ebook formats, DRM FREE
- Participation in the O'Reilly community
- Newsletters
- Account management
- 100% Satisfaction Guarantee

Signing up is easy:

1. **Go to: oreilly.com/go/register**
2. **Create an O'Reilly login.**
3. **Provide your address.**
4. **Register your books.**

Note: English-language books only

To order books online:

oreilly.com/store

For questions about products or an order:

orders@oreilly.com

To sign up to get topic-specific email announcements and/or news about upcoming books, conferences, special offers, and new technologies:

elists@oreilly.com

For technical questions about book content:

booktech@oreilly.com

To submit new book proposals to our editors:

proposals@oreilly.com

O'Reilly books are available in multiple DRM-free ebook formats. For more information:

oreilly.com/ebooks

O'REILLY®

CPSIA information can be obtained at www.ICGtesting.com
Printed in the USA
BVOW030548210612

293282BV00002B/1/P